Evolution's Final Days

The Mounting Evidence Disproving the Theory of Evolution

By John Morrison

Copyright© 2019
All Rights Reserved
ISBN: 9781091497764
Published by ZML Corp LLC

TABLE OF CONTENTS

Disclaimer..3

About the Author..5

Introduction 7

Part 1: Microevolution..11

Part 2: Macroevolution 15

Part 3: Dual Sexes 19

Part 4: 2nd Law of Thermodynamics 25

Part 5: Irreducible Complexity 29

Part 6: Humans Are 99% Identical to Apes...........41

Part 7: The Fossil Record 45

Part 8: Questioning Science....................................53

Part 9: Origin of Life 59

Part 10: Effects of Mutation....................................63

Conclusion..71

DISCLAIMER

This book is written for informational and entertainment purposes only. The author and publisher are not affiliated with any schools, organizations, or people mentioned in this book. It is illegal to copy or distribute any part of this book without written consent from the author or publisher. Written by John Morrison, published by ZML Corp LLC. Copyright © 2019, all rights reserved.

This Page Intentionally Left Blank

About The Author

◆ ◆ ◆

My name is John Morrison. I was born in Newburgh, New York, United States and raised in a Christian household. I classified myself as a Christian until the end of my college years when I started shying away from organized religion. As I studied more, I began to find myself claiming to be "spiritual" as many would say, and not a part of any particular denomination.

I do not consider myself a "creationist." I do not believe the earth was created in 7 days or Adam and Eve were the first humans on earth. I'm quite analytical, and my family often tells me I would have made a great lawyer. I look at the facts and evidence presented in an argument before drawing conclusions and making a decision. Ever since I first started learning about evolution I've had issues with it. Logical facts are not there, but we as a society are told by the scientific community it's how life was formed. While delving deeper into the theory during college, I quickly learned students were supposed be quiet, not ask questions, and accept the theory. I would compare it to when the government told us going to war in Vietnam would be good for the country, and to just accept their advice and support the war. If a good theory was presented by scientists that clearly showed how life was formed, I would be more than open to it. But to present something as a theory that has so many gaps and holes and not allow for alternative suggestions is never how science was intended to be.

In many high school and college classes, evolution is taught. Whether it's a summary or a more in depth discussion, many parts are skipped over or not mentioned. Why is this? It's because they have no explanation for them. In this novel, I will go over the many holes in the theory of evolution which biology textbooks fail to mention. I will show you how the world gravitates towards disorder, and why the theory of evolution doesn't fit with the framework of our planet. I will talk about the scientific community, and why questioning evolution results in ridicule and ex-communication. And I will prove to you that evolution as presented by scientists today could have never happened, and is not a viable theory to explain how life on this planet started.

Introduction

◆ ◆ ◆

For thousands of years, humans have been coming up with ideas trying to explain how our world operates and how we got here. We are fascinated by the complexity and intricacies of our world and have a lust to explain it. Before the 1800's, there weren't many good ideas to explain the existence of humans. Many scientists held Christian-Judeo beliefs, and debated on when God created humanity and not how it actually started. Scientists at the time believed life forms were fixed, and everything was created as it was, as in nothing changes. In the early 1800's, a group of scientists began quietly questioning this notion. Jean-Baptiste Lamarck is considered the first evolutionist, and briefly predated Darwin. He held a belief known as "soft inheritance," or use and disuse. This belief stated that organisms do change over time, and this change was related to adaptations while living, which was passed onto offspring. An example being a giraffe's neck muscle was strengthened by trying to reach the top of a tree, thus making its neck grow longer. And over many generations, as each giraffe tried to reach the top of tree branches, their necks grew more and more each time which was then passed down to their offspring. This is obviously not true, as we now understand genetics doesn't work this way. For example, if you were to practice playing basketball every day and become the number one basketball player in the world, this learned trait is not passed down to your offspring.

The next explanation presented for how life formed was that large events, such as a meteorite hitting the earth, caused the variation we see in the world. This type of event would have caused rapid change, which would have quickly changed the species present on earth. This theory drew criticism and was not widely supported in the scientific community.

Then in 1859, Charles Darwin published his book titled "On the Origin of Species." In this book, he suggested the idea of slow gradual change by natural selection over many millions of years was the explanation for the diversity of life on the planet. He suggested that humans were just the descendants of a long line of other species which had morphed, or evolved, over a long period of time. The scientific community soon widely accepted his viewpoint, and eventually the theory of evolution was born. Something to note is just like Lamarck, Charles Darwin believed in "soft inheritance." As in an animal's traits could be passed on to future generations simply through acquiring them in this life. Darwin did not have an understanding of DNA or genetics like we do today.

Another item I want to point out is the difference between a hypothesis and theory as defined by the scientific community. A hypothesis is a suggested explanation for a phenomenon. A theory is a well-tested, well-substantiated, unifying explanation for a set of proven factors. A theory has been tested and is backed by evidence. Evolution is currently considered a theory in the scientific community even though it technically does not really meet this criteria, as I'll explain in this novel.

PART 1
Microevolution

◆ ◆ ◆

There is no question that organisms change over time. This is a known fact and not even the most hard core creationist would argue it. However there are two types of evolutionary processes: microevolution and macroevolution. Microevolution would be considered changes within a species, usually over a short time period. A good example of this would be bacteria becoming resistant to antibiotics. Penicillin was developed in the late 1920's, and soon became widely prescribed to treat infections. Over a relatively short period of time, certain bacteria were able to develop resistance to penicillin and could even multiply in the presence of it. Thus, the bacteria had gone through changes in its makeup, or evolved, for this to be possible. Another example is the evidence biological textbooks famously consider to be proof of evolution; the changes in finches beaks on the Galapagos Islands. For those unfamiliar, in the 1800's Charles Darwin visited the Galapagos Islands, and noticed that the finches on each of the islands varied slightly, particularly in their beak. This observation piggy backed on the ideas brought forth by Lamarck, and helped Darwin develop his explanation for life in the form of evolution by natural selection. The finches on each of the islands had evolved over many generations to have the best shaped beak for their particular island, and the finches with poorly shaped beaks died off and no longer produced offspring. Thus, natural selection caused the evolution of the finch species in the Galapagos.

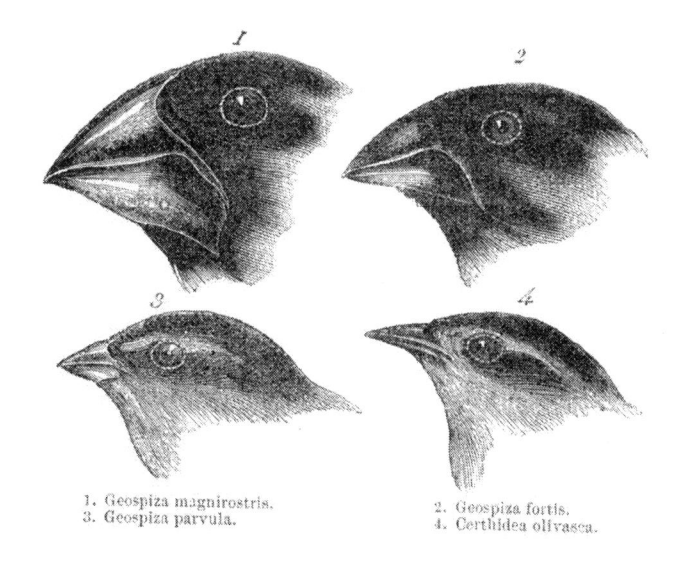

1. Geospiza magnirostris.
3. Geospiza parvula.
2. Geospiza fortis.
4. Certhidea olivasca.

I first want to point out that no one is denying that natural selection is a reasonable theory and does take place on earth. But I want you to notice something here which is not mentioned in biological textbooks when they mention this "proof" of evolution; this is microevolution! The finch did not become a fish and move off into the ocean. Simply, over many years, the finches with the better suited traits survived, and finches with traits less suited for their environment died off.

While microevolution is a part of the theory of evolution, macroevolution is an even bigger part and if not true, would discredit the entire theory.

As a **token of appreciation** to my readers, I am offering my special report above titled "*Top 5 World Mysteries*" **absolutely free**! Just copy and paste the link below into your browser, put in your email address, and it will be immediately sent to you.

fastlink.xyz/topfive

PART 2
Macroevolution

◆ ◆ ◆

Macroevolution is defined as the evolution of groups outside of their own species. An example would be the evolutionary claim water dwelling animals like reptiles evolved into land dwelling mammals. Macroevolution has never been observed and for good reason; macroevolution is supposed to be a very gradual process, taking place over millions of years. Considering this, the actual observation of it cannot take place, which evolutionist point out whenever the question is brought up. However, had there been such transition from species to species taking place over an imaginary time period evolutionist cite and is continuing to this day, wouldn't there still be species transitioning? Why has the transitioning of species stopped? There is absolutely no scientific evidence of crosses between species. Let's look at an evolutionist's chart showing the tree of life.

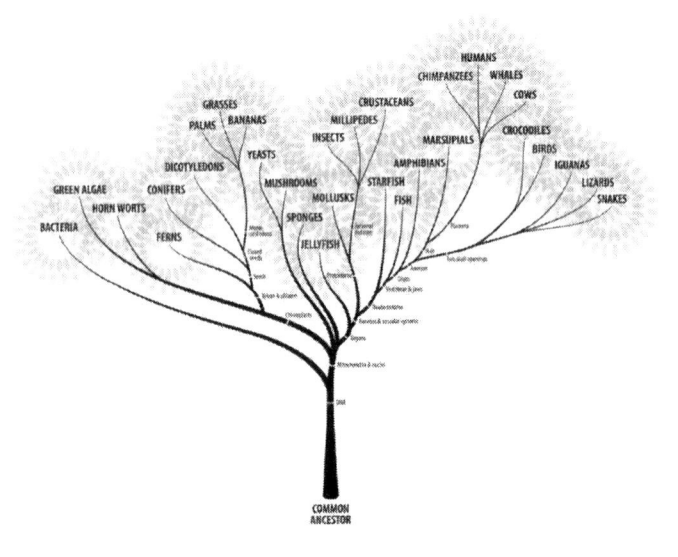

As you can see from the chart above, the evolutionary theory states we started from a common ancestor (single celled organism), which soon began breaking off into different families. And from those points, more distinct species were formed. While it's a nice looking chart, the problem we run into is it's not what we observe here on earth. Neither our own observation of currently living species nor the fossil record show any indication of such a chart. Instead, this is what we see.

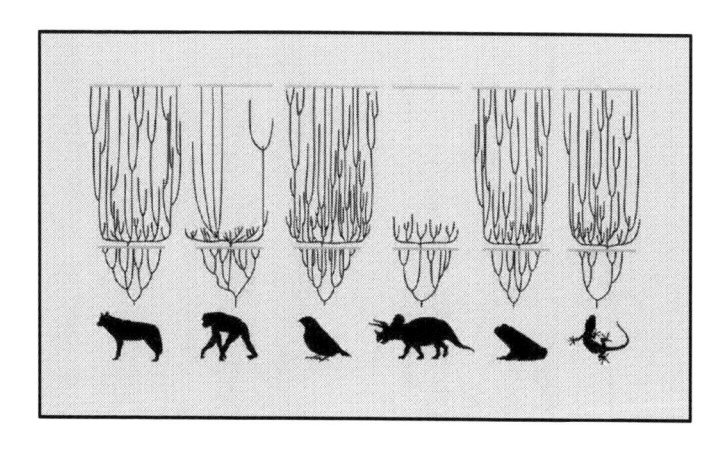

The chart above is what we actually observe in the world around us. There are no "transitional" species, but instead one species in different forms. An example would be a cat. You have a cheetah, a lion, a house cat, etc. While different, they all belong to the same family of animal called Felidae.

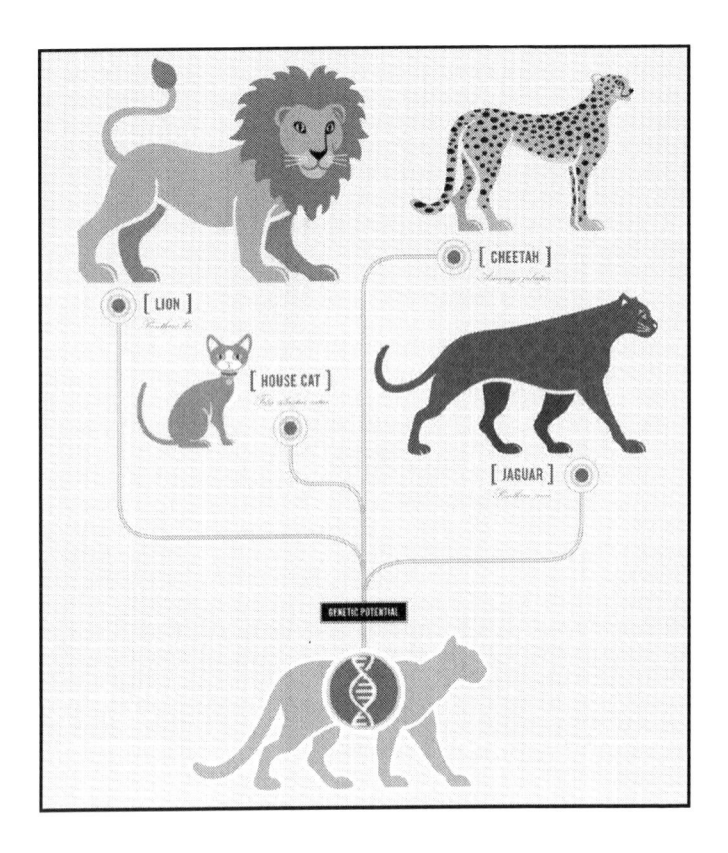

This is another example of microevolution, as opposed to macroevolution. And the fossil record, which we will touch on later, doesn't do much for evolutionary theory either. It shows the same exact types of formations as we see in current times, without the "transitioning" species that evolutionists claim to have existed.

PART 3
Dual Sexes

◆ ◆ ◆

This conundrum in the evolutionary theory is completely skipped over in most college textbooks. You may have not even heard of this issue before, but it's important to bring up here.

While hard to believe, scientists stand firm that all life stems from a single celled organism, which evolved over billions of years into the life forms on earth today. While this could be somewhat plausible for an asexual organism, having two sexes creates problems in the theory. Let's look at the human penis and vagina. The two parts of the human body are meant for each other. A man becomes aroused when in the presence of a female, or even just thinking about a female. Then his penis fills with blood and becomes erect, allowing it to be inserted into the vagina. The vagina, being the correct size for the penis, then becomes moist allowing for easier penetration. An orgasm then occurs in both men and women, which begins of the process of life. Based on what we know, there was never an asexual human. This would mean that not only would a man and woman have to have evolved by themselves, they would also have had to co-evolve together. Considering the incredibly small probability of having one entity evolve, what would be the chances of two entities CO-EVOLVING next to each other at the same time, and being perfect for each other? What would be the evolutionary advantage of moving away from asexuality and needing another member of your species to mate with? There is a lot of work involved in finding a mate, potential issues in reproduction, etc. that would have no basis in efficient evolution.

Humans are said to have evolved from a distant ape relative. And apes are said to have evolved from another mammal, which ultimately all started somewhere in the ocean. The penis and vagina are common in many animals, including ocean dwelling ones. For this reason, I can understand if evolutionists say they just came with each species into the next evolutionary phase. The issue is the start of it all. How are we supposed to believe that some animal in the ocean which was asexual had a mishap in its DNA to form a penis, and another had a mishap to form a vagina? And then the two started having intercourse with each other? And one developed a vagina with ovaries and another a penis and testicles to start a two sex world. And the penis and vagina were able to fit into each other and be pleasurable at the same time? There are a lot of "how in the heck" questions here. You will hear a lot of jibber jabber from the evolutionary community but I want you to think with your own mind and try and make sense of this logically… it doesn't make sense!

Something else to note is sperm is kept in the testicles. In humans, the testicles are held outside of the body, because they have to be kept at a precise temperature or else the sperm die. Somehow male apes had to evolve to have their testicles outside of their body BEFORE they were held inside the body. But how is this possible? While I'm not a medical doctor and there may be more examples like this, this is just one example to show the precision in the body needed to create life. And without it, life couldn't have existed.

Another big issue is that of mitosis and meiosis. If you've taken a biology class, you may remember these processes in cells. Cells can divide and replicate in two ways: mitosis and meiosis. Simple, asexual organisms, go through mitosis. These would have been the organisms which dominated early earth, according to evolutionists. In mitosis, one cell divides into two identical daughter cells, which contain the same genetic information and the same number of chromosomes as the original. Meiosis is much different. Meiosis occurs in all sexually reproducing organisms, as it is required in order to make offspring. In humans, meiosis is the process which occurs in sperm cells and egg cells. During meiosis, one sperm or egg cell divides into four daughter cells, each with different characteristics and half the number of chromosomes of the original cell. Then, during intercourse, one sperm cell with only 23 chromosomes fuses with an egg cell with only 23 chromosomes, to make a full 46 chromosome human cell. This cell then replicates via mitosis, and eventually a complete human is made. With all this being said, the big issues here are how and why?

How is this possible? Like other aspects of organisms we'll talk about in this book, changing from mitosis to meiosis would have been an extremely complicated process. Meiosis has been coined the "queen of problems" in evolutionary biology, as to this day, a reasonable explanation for how it would have been adapted by an asexual organism has not been given by scientists.

And another question is why? Why would a perfectly functioning, asexual organism which is able to reproduce on its own just fine via mitosis, go through the pain staking process of evolving to have cells which undergo meiosis? What would be the evolutionary advantage to this? None of said organisms peers would be able to make use of this evolutionary anomaly, and there is no reason to postulate this anomaly would have continued on.

PART 4
2nd Law of Thermodynamics

◆ ◆ ◆

There are four accepted laws of thermodynamics. These are the laws scientists have agreed upon involving fundamental physical quantities which include temperature, energy, and entropy. These laws are applicable to all scientific fields which include chemistry, physics, and biology.

The 2nd law of thermodynamics states that over time entropy, also known as disorder, increases. In the scientific community, entropy is shown as ΔS (delta S). In nature, molecules gravitate towards randomness and disorder. Over time, scientists observe processes on earth become MORE disordered, not less. Let's take for example your high school yearbook. As the years go by, you notice your high school yearbook becomes more unstable. The pages start deteriorating, the pictures start fading, and the bindings slowly start coming unglued. Another example is computers; over time, computers become less reliable, crashing more, having hardware issues, malfunctions, etc. In the human body, aging is an example of the 2nd law of thermodynamics.

Let me show you an example of gas molecules in a container demonstrating increasing entropy.

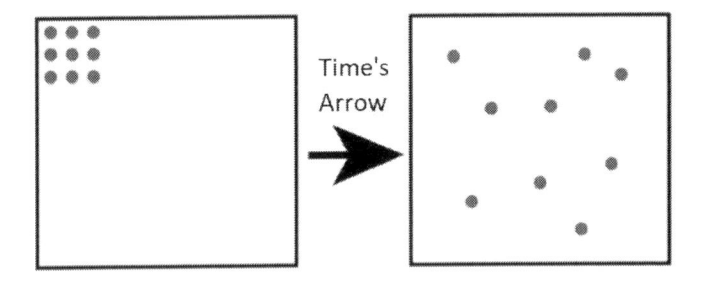

The arrow represents time's effect on the molecules. With time, they do not come together and start building on each. Instead, they start spreading out, becoming more random and distant, and eventually they completely disperse.

Here is another example. Let's say you threw a pile of bricks out of a truck. Which picture below would be more likely to be the outcome?

Disorder is more probable than order.

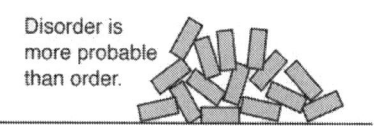

Obviously the picture on the right would be the more expected outcome. And this is because of the 2nd law of thermodynamics which states the universe moves towards disorder. You could throw that same pile of bricks out of the truck 1,000 times (or for 300 million years if you're an evolutionist) and never get a picture that even comes close to the first pile. The only way to do that would be to actually design the bricks, which we will touch on later. So why would this randomness and disorder apply to every field of science, including biology, but somehow by exempt from evolution?

The theory of evolution states that random amino acids somehow came together, thus violating the 2nd law of thermodynamics and became a cell. Then, this single celled, asexual organism randomly mutated and again became MORE ordered and has since become more ordered with every mutation. Their explanation for this is it's been millions and millions of years, an imaginary time period that makes their theory appear more believable because of how long of a timeline they paint. I don't care how long it's been, life does not become more ordered with time, it becomes more disordered. Now that's not to say it doesn't change with time, as we see with microevolution. However to become more ordered, such as the ability to start seeing, hearing, feeling, etc. is an absolute stretch of the imagination.

Let's go back to that pile of bricks you threw out of the truck. If you were to leave them in a pile for 5 million years, what would you suppose would happen to them? They would not form a perfect ordered stack, and grow arms and start moving. They would eventually degrade and spread out further and further, becoming even more disordered than when you threw them out of the truck. And had you broken the laws of probability and they did land in a perfect, ordered pile as shown on the picture on the left, they would still start to degrade and spread out further over time. And evolutionists can replace that pile of bricks with some amino acids, as we will talk about in part 9 of this book, and the same outcome would result. No part of science is exempt from the 2nd law of thermodynamics, and that includes evolution.

PART 5
Irreducible Complexity

♦ ♦ ♦

Sir Isaac Newton, one of the most distinguished scientists in history once said that the physical laws he had uncovered,

"revealed the mechanical perfection of the workings of the universe to be akin to a watch, wherein the watchmaker is God."

This analogy was brought into the mainstream when scientist William Paley published a book detailing it in 1802. The argument states that you could compare a watch to life forms here on earth. When you look at a watch, it has many different moving parts and components. If you were to be walking around a park and found a watch on the ground, you could clearly tell that it was built by someone. If you were argue that it was always there, and formed naturally over millions of years, others would call you crazy. The watch clearly came from a designer; someone made the watch. Why is it any different with human life?

Humans, as well as the many other life forms on earth, have complex, intricate systems. These intricate systems point to some type of designer, and to argue they formed naturally over many years is as crazy as saying a watch could do the same. Not only this, the watch is made up of many different parts, some of which would be obsolete without the others. The motor wouldn't turn without the battery, the watch hand wouldn't turn without the motor, etc. This is known as something called "irreducible complexity," meaning that without all the parts of the system combined as one, the system becomes useless. Irreducible complexity is highlighted in many biological organisms here on earth. The main premise behind irreducible complexity is if something is taken out of the system, the system fails to continue working. This would mean the system could not have been built in successive modifications, as everything is needed at once in order to make it function properly. It's all or nothing! Either the everything is there and the system works or something is missing and it doesn't; there is no in-between!

Charles Darwin is quoted saying in his book, On the Origin of Species,

"If it could be demonstrated that any complex organ existed which could not possibly have been formed by numerous, successive, slight modifications, my theory would absolutely break down."

This means that, based on what the founder of evolution himself stated, if something existed here on earth which could not be explained by successive modifications over time, it would disprove the theory of evolution. And, based on what we're about to go over, this would mean the theory has been disproven.

While there are many examples of irreducible complexity in the biochemistry of humans, I will only be touching on just a few examples in this book. Many times in evolutionary biology, these hard to cover topics are just briefly touched on. When questioned, poor explanations are given on how they formed. But once you dig a little deeper, the sheer engineering marvel of the biochemistry and complexity of these structures and processes make little sense from an evolutionary standpoint. Just like the watch, without all the parts coming together as one, each part by itself becomes meaningless. And not only this, let's say hypothetically all the parts were there and ready to be used; they would have to come together perfectly to form a working structure. And as we learned from our pile of bricks in Part 4, this is impossible based on our universe's 2nd law of thermodynamics.

The Eye

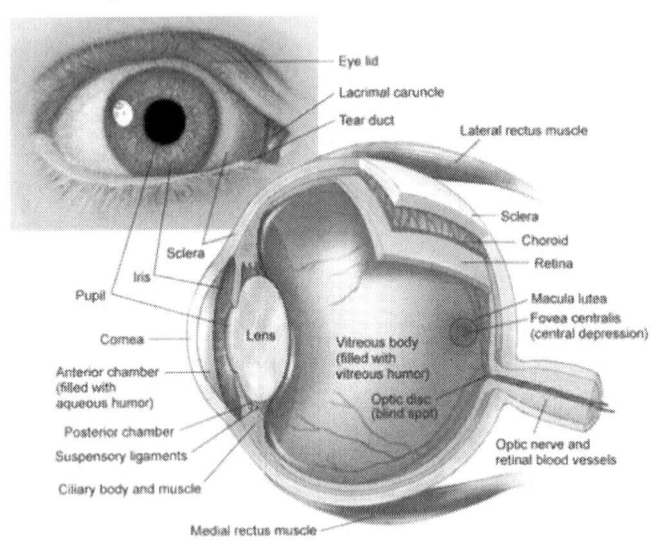

As the picture above shows, there are many parts to the eye which allow it to function properly. Evolutionists claim that a random mutation occurred which allowed one lucky organism to have a sensitivity to light, and then over time this light sensitivity developed into an eye. And then that eye became more developed and eventually evolved into what it is today. There are a couple issues with this supposed explanation.

The first issue would be our observation of animals, both here presently and in the fossil record. There is no transitionary eye in an organism. In the fossil record the eye just appears and it appears as a very complicated structure. The earliest known fossil of an eye is dated 540 million years ago, which is the Cambrian era. It is that of a trilobite, which is a type of beetle. This isn't to say there aren't different types of eyes. Different species have different types of eyes, ranging in size, shape, sensitivity, etc. This variety can be explained with natural selection and microevolution, which no one is denying. The issue is how this eye came into place. The fossil record, which we will touch on more later in the book, shows eyes evolving independently and not all coming from one common ancestor. This refutes the evolutionary explanation of how one common ancestor gave all organisms on earth vision. One eye evolving is hard to believe, but 20 evolving individually? 30? The odds of something like this are not even possible.

But let's just say this did happen and it all started with a light sensitive area. What good is a light sensitive spot on your body without it connecting to your information processor, AKA the brain? Did it just sit there on the organism until an optic nerve connected on to it, as well as on to the brain? It would make little evolutionary sense for it to sit there considering it has no use. That light sensitive area is useless unless it's connected to the brain where the organism could make use of it. And did this process happen 30 times with each organism that evolved a light sensitive area? We're not making much sense at this point.

The second issue is the irreducible complexity of the eye. Without all the parts of the eye coming together as one, it ceases to function. Evolutionary theory states gradual changes occur over time, each one building on the other. The complexity of not just the eye, but even just a light sensitive area is extraordinary. Biochemist Michael Behe, referring to just a "light sensitive area" states, "each of its cells makes the complexity of a motorcycle or television set look paltry in comparison." Think about a motorcycle or a television set for a second and how complex it is. And we're talking about just a light sensitive area here, not an actual eye! Could something this complex come together naturally? Not a chance.

In the rudimentary eyes, there would have also had to be a "transparent layer" which we know as the cornea. This transparent layer requires a very well organized structure of corneal fibers, and these fibers need a chemical pump to make sure they contain exactly the right amount of water going into them. Are you starting to see the irreducible complexity? And we're still just at a light sensitive area, not a fully formed eye yet.

Biology textbooks just say things such as "the process happened slowly" without a detailed explanation. The reason why is they don't have one! This process can't happen. This type of process, even just a "light sensitive area," has complex beginnings that cannot be explained by simple building blocks on top of one another.

Bacteria Flagellum

While the world we live has its own complexities, the microscopic world is even more complex. Cells contain an organelle called the flagella. An organelle is just a structure within a cell which operates like an organ (liver, kidney, etc.), having a specific task. The flagella is the primary method of transportation for cells. Below is a picture of a few different bacteria with the flagella attached to the end of them. It resembles a tail you would see on an animal.

When we zoom in to where the flagella attaches onto bacteria, we discover a tiny motor (shown on the next page) which is the driving force behind the flagella. It has been touted as "the most efficient machine in the universe."

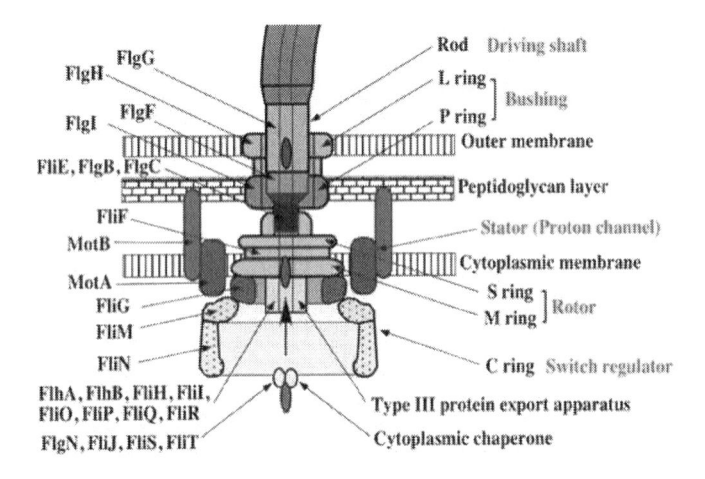

Diagram of flagella motor with labeled parts: FlgG, FlgH, FlgI, FlgF, FliE, FlgB, FlgC, FliF, MotB, MotA, FliG, FliM, FliN, FlhA, FlhB, FliH, FliI, FliO, FliP, FliQ, FliR, FlgN, FliJ, FliS, FliT; Rod (Driving shaft), L ring, P ring (Bushing), Outer membrane, Peptidoglycan layer, Stator (Proton channel), Cytoplasmic membrane, S ring, M ring (Rotor), C ring (Switch regulator), Type III protein export apparatus, Cytoplasmic chaperone.

To function correctly, this flagella motor must contain over 40 different proteins, which are labeled on the picture above. Remove just one of these 40 proteins, and the flagella stops working. Similar to a watch, it needs all its parts, and if you take away just one, it fails to function.

You may be thinking that's fine and dandy John, but all those processes could have just built off of each other, thus validating the theory of evolution. The problem is they couldn't. While some of the components of the flagella motor can be explained by co-option, as in they were used by other parts of the cell and then over taken and used by the flagella, most of the flagella is considered "new." This would mean it doesn't appear in other parts of the cell. And like blood clotting which we will talk about next, it needs all of it's components to operate. Without every single one, the flagella is useless.

Not only this, if the universe got flipped upside down and all this components somehow appeared at once, they would still have to assemble into a perfect working machine which is not possible! That's like claiming if you have all the components to a car motor in your garage, such as the pistons, rods, casing, etc. they could just fall off the bench and assemble into a perfectly working engine. Obviously this could never happen and the 2nd law of thermodynamics helps proves this. As you are finding out, the 2nd law of thermodynamics is becoming a very important part of this book

Blood Clotting

When you get a cut, you start bleeding. The size of the wound can determine the length of the process, however soon the bleeding slows and eventually comes to a stop. This is a process known as blood clotting. It's something we don't even give a second thought to, but without it blood would continuously flow out of our body leading to certain death. There are a few extraordinary factors associated with blood clotting. One is the fact the body somehow knows it is bleeding and is able to determine the exact length, width, and location of the wound to clot the blood. If the blood were clot at the wrong place in the body, it could block circulation, resulting in a heart attack or stroke. If the blood clot were not confined to just the wound, the entire blood system could clot, causing death.

There are a total of 12 factors associated in the blood clotting of humans. As in 12 processes that piggyback off of each other, allowing the next process in the series to work effectively. And you can't just start with the first, and then naturally evolve over millions of years to the 12th, as all twelve are needed for clotting to work properly. A study was done with mice by a professor at UC San Diego by the name of Russell Doolittle. He found that mice were able to survive with two blood clotting factors (plasminogen and fibrinogen) eliminated. That's quite novel from an evolutionary standpoint. The issue is what happened to the mice when these blood clotting factors were removed.

The mice with just plasminogen removed suffered from "uncleared blood clots." This means the blood would clot, but wouldn't be absorbed into the body. And this led to a blockage in the blood, which caused a host of health issues and ultimately death. Mice with both the plasminogen and fibrinogen removed, didn't have this issue. That is because they had no clotting function at all! So sure they lived just fine, but any sort of injury which resulted in a cut meant bleeding out indefinitely until they died. So while trying to cite reasons to validate evolution, this professor effectively invalidated his own theory with this experiment!

Human Body

While any mammal could be elaborated on, I'd like to focus on the human body, as it's what we can most relate to. You don't need to be an anatomy professor to realize how complicated the human body is. The organ system in particular is a complex, interconnected system in the human body. We have many different organs which together, allow us to live. The heart pumps your blood, the brain sends signals throughout your body, the kidneys filter your blood, etc. How did this system "evolve" into place? If just one of your organs malfunctions or is removed (except for half a kidney or an appendix), the human body fails to operate and we die. The human body itself is the epitome of irreducible complexity.

Many times you may hear evolutionists bring up "vestigial organs." These are "useless" organs, which "serve no purpose" and are still inside the body because they were never dissociated from past evolutionary ancestors. A prime example the evolutionary community has pointed to in the human body is the appendix. The issue with the appendix is we really didn't know what it did, so evolutionists automatically assumed it was a "vestigial organ."

This in turn helped to perpetuate their theory, as if we have "useless organs," there is no intelligent designer. While we have made large strides in medical discoveries over the past century, we still don't know everything.

As we are now discovering, the appendix is not a "vestigial organ," and does serve a purpose. Studies have found it to be involved in storing good bacteria and making white blood cells. Long term studies have also shown there are potential negative side effects of people who have had their appendix removed. This all goes to show that just because we have not yet discovered the purpose of something does not mean it's "useless."

What's funny is complexity in all other areas of study except for biology is considered evidence of design. Look at reports of alien visitors when elaborate crop circles are discovered, finding prehistoric tools in the ground, or stumbling upon ancient monuments like Stonehenge. You would be laughed at in the scientific community if you proposed something as complex as Stonehenge was produced naturally over millions of years from wind, rain, and a possible electric spark. However this is exactly what the theory of evolution proposed happened with living organisms (which are infinitely more complex). Quite a dilemma we have here.

PART 6
Humans Are 99% Identical to Apes

♦ ♦ ♦

The first thing I'd like to mention here is our DNA is also said to be 60% identical to a banana, and 80% identical to a dog, both of which are fairly different than human beings. I'm sure you've heard this statistic a few times in your lifetime. We as humans came from apes, and it's proven, as our DNA is 99% identical to their own. Do you know where this statement comes from though?

Genes are made up of DNA. DNA is made up of nitrogenous base pairs, which twist into a double helix structure or twisted "ladder." When unwound, each "step" of the DNA ladder is a base pair.

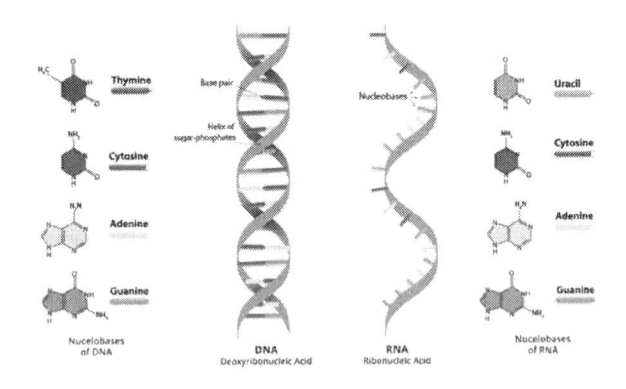

A base pair is made up of two nitrogenous bases. These bases include adenine (A), guanine (G), cytosine (C) and thymine (T). Each of these bases can only pair with one other base (A with T, G with C). Knowing this, if only one strand of a DNA helix is shown, the other side can be calculated. An organism's "genome" is the sum total of these base pairs in its cell put together.

In recent years, scientists have been able to map out the entire human and ape genome. Say you take apart the DNA, so you have one side (as shown in picture below), and then lay it out flat onto a piece of paper. We would see all the nitrogenous bases, which we will label as letters.

AGTCCGCGAATACAGGCTCGGT

When writing an essay, many letters are contained together, which would be called a paragraph. In our genome, we'll call this paragraph a sequence. When we compare the DNA genomes of apes and humans, there are single letter differences in these sequences which can easily be accounted for. But then there are differences that aren't so easily accounted for. They include:

- *Genetic sequences thousands of letters long contained twice in the human genome, but only once in the ape genome.*
- *Identical sequences in different places*
- *Identical sequences in reverse order*
- *A sequence in human DNA that is broken into pieces and occurring in different areas of an ape's DNA.*

Considering these "minor annoyances," researchers comparing the two genomes just excluded these mismatched sections. This means they excluded 1.3 billion letters of DNA! They then performed a letter by letter comparison on the remaining 2.4 billion letters in the genomes, which turned out to be 98.77% identical. So yes, we share a fair amount of DNA with apes… if we completely ignore 18% of an ape's genome and 25% of ours. This is the epitome of "fudging" data to fit your theory and to say we're "98.77%" identical to apes is an inaccurate statistic and not valid scientific research. That's like saying a car is 98.77% identical to a boat if you take away the paint color, transmission, wheels, brakes, etc. and just include the engine and steering wheel. Sure, they're exactly alike!

Another issue with this statistic is more mutations in DNA doesn't always mean bigger changes. Sometimes just a few changes in DNA can produce big changes, where a large number of mutations could produce minor changes. It depends on what part of the DNA is being changed. This means just comparing the percentage difference of DNA between two organisms doesn't tell us everything, and is just a trick to appear like a bigger deal to the uninformed.

PART 7
The Fossil Record

♦ ♦ ♦

The fossil record is the most reliable form of evidence, as it's a window into the history of the organisms that have lived on this planet. It also is one of the most compelling arguments against evolution. Darwin is quoted saying:

"But just in proportion as this process of extermination has acted on an enormous scale, so must the number of intermediate varieties, which have formerly existed, be truly enormous. Why then is not every geological formation and every stratum full of such intermediate links? Geology assuredly does not reveal any such finely graduated organic chain; and this, perhaps, is the most obvious and serious objection

which can be urged against the theory..."

So Charles Darwin, the originator of the theory of evolution is quoted saying that the incomplete fossil record puts a hole in his theory. And since Darwin's time on earth in the 1800's, evidence for evolution in the fossil record hasn't improved, and has actually painted a more detrimental picture for the theory.

The fossil record shows low complexity, asexual lifeforms over many millions of years and then BAM... complex organisms. This mass explosion of life is called the "Cambrian Explosion" because it happened during the Cambrian era of the fossil record. Now again, I want you to think logically with a reasonable mind and not listen to evolutionist jibber jabber. If evolution were real and very gradual changes happened over millions of years, wouldn't there be fossils in the ground to show this change? The explanations by the evolutionary community make little sense and they really don't have a plausible explanation for it. For the most part they sweep it under the rug, along with everything else that goes against their theory.

One of the arguments evolutionists present for why there are so few pre-Cambrian era fossils found is because of the possibility that many of these creatures were "soft bodied" and thus, couldn't form into a fossil. Darwin is quoted saying "No organism wholly soft can be preserved." This however, is not true. Many "soft bodied" fossils such as jellyfish and worms have been found in the fossil record. Archeologists have even discovered fossils of sand ripples and rain drops, and these aren't even organisms!

And let's step away from the Cambrian explosion for a moment, and go to present day life forms. As mentioned in part 2 of this book (macroevolution), not only have no fossils been found which show the transitional phase, there are no life forms currently living in this phase. Evolution is purported as extremely gradual changes over many millions of years. This means there should be thousands and thousands of transitionary life forms which appear in the fossil record, as well as organisms in transition currently living on earth, which show this gradual change in macroevolution. We see none… zero. Not even one. How can you explain this? This issue in the fossil record is by far the most damaging evidence against evolution there is.

The best evolutionists can give us to prove humans evolved from apes is a fossil they discovered named Lucy. Now let's again think for a minute. Millions of years of transitioning species, millions of species who have lived and died, very gradual changes, and only one very questionable fossil can be found that kind of looks like an ape-human hybrid? That doesn't make much sense.

Lucy was a fossil discovered in the 1970's by a

group of researchers on the continent of Africa. The fossil adds credence to evolution, as it supposedly shows a human/ape hybrid which walked upright, and is the missing link scientists have been desperately searching for. There has been a myriad of debate surrounding this fossil. Some scientists say it walked upright, others say it was a knuckle walker, others say the fossil found is too incomplete to form an opinion. I have a picture below of the entire fossil that was found.

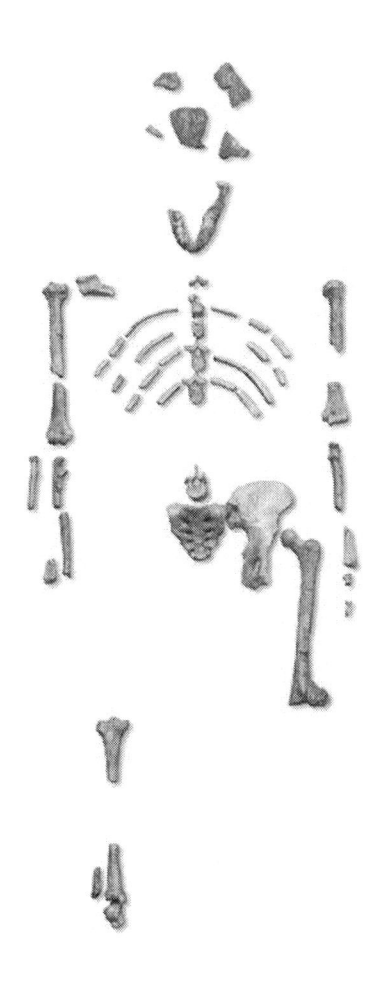

Notice much? I don't either. No skull, no hands, no feet, and then bits and pieces of other parts of the body (it was later found that one of the ribs belonged to a baboon). Somehow though, this incomplete fossil has been turned into complete replicas of what the ape would have looked like had it been living today. Below is a model shown at the St. Louis Zoo in Missouri.

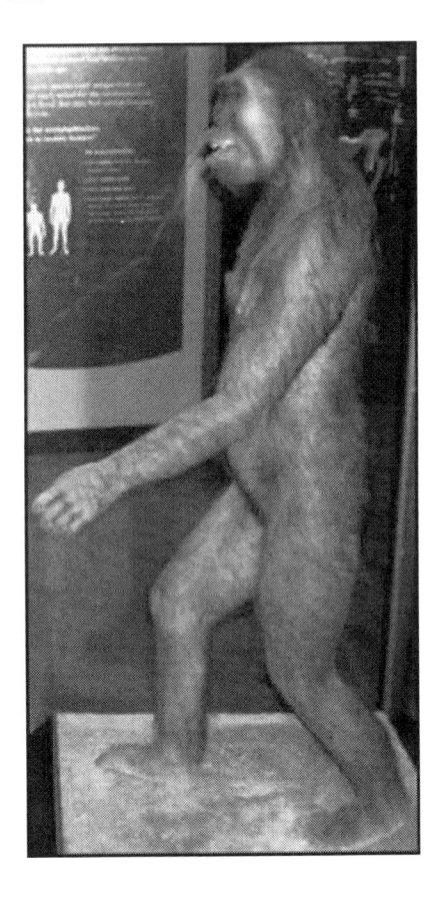

The model displayed at this zoo, and in many other museums around the country, show a fully formed ape/human hybrid standing upright on two legs. This model was somehow extrapolated from the incomplete bone fragments (shown before) of the Lucy fossil. As you can see in the picture, a fully formed gorilla-human hybrid with a face, feet, and hands is shown, none of which were a part of the fossil fragments found. They extended their own opinions of what this fossil "should" have looked like based on their own biases. Another common picture, shown below, is what the face of Lucy "should" have looked like, had it been alive today.

Pictures like this are found in many biology textbooks when explaining human evolution, and the discovery of the Lucy fossil. This is what is taught to the high school and college students of America, thus skewing their opinion with unfounded pictorial evidence. I find this to be quite deceiving. What they should be showing in textbooks and museums is the actual incomplete fossil found and allowing the public and students to decide for themselves. This practice of showing a fully formed, computer simulated model of "Lucy," and stating this is what was found and moving on to the next subject is obviously deceitful. It shows the practice of the evolutionary community as being untrustworthy in their evidential approach. They should also explain the ongoing debate in the scientific community over whether or not this was actually an ancient ancestor or just an oddly formed ape, instead of stating complete consensus of the former.

PART 8
Questioning Science

♦ ♦ ♦

Science is a fascinating endeavor that has helped drive countless innovations to improve life here on planet earth. Ever since the concept of science formed, there have been scientists coming up with hypotheses and theories explaining how things here on earth work, as well as other scientists coming up with competing hypotheses and theories. The scientific community openly accepts competing theories which offer a better explanation than the current theory in place. This is what has allowed science to progress and innovation to continue throughout history. Unfortunately, this is not the current scenario with evolution.

Evolution is the one and only theory for how life on earth was formed that the scientific community promotes. There is no other theory even competing for space on the subject, and mentioning another theory is a sure fire way to get ex-communicated from the scientific community.

A Chinese paleontologist by the name of Jun-Yuan Chen is famously quoted as saying,

"In China we can criticize Darwin, but not the government; in America, you can criticize the government, but not Darwin."

There are really only two other explanations which are currently able to explain life on earth besides evolution. The first would be creationism, as taught in many world religions. And the second would be intelligent design. Many would link intelligent design to a Christian-Judeo God, but it doesn't have to be. What if aliens intelligently designed us? Or our collective consciousness? Couldn't this be reasonable in the scientific community? Even the notable scientist Stephen Hawking is quoted saying,

"I believe alien life is quite common in the universe..."

This is much more believable than some currently accepted theories in the scientific community with one example being the "multiple universe" theory, which states there are an infinite number of multiple universes existing side-by-side ours, each with their own timeline. While it seems far-fetched, this example is openly accepted as a "theory" in the scientific community. However, a scientist with a reasonable explanation for how life formed on earth cannot even mention an alternative theory such as intelligent design (which actually proves to be a better explanation for life than evolution) without being slandered and ridiculed. Why is this the case? Let me go over some examples of professors and lecturers in academia who have tried to present alternative theories and the consequences they endured.

Caroline Crocker was a lecturer at George Mason University in Virginia. Caroline states she mentioned intelligent design in her cell biology class, and her academic career suddenly came to a halt. She was fired from her position at George Mason and blacklisted in academia. After this blacklisting, she was no longer even considered for positions at other institutions, where in the past she was offered jobs on the spot after an interview.

Michael Egnor is a neurosurgery professor at Stony Brook University in New York. A high school in Virginia was having a contest with the topic "Why I would want my doctor to have studied evolution." Michael wrote a post on an intelligent design website which stated evolution was irrelevant to medicine. Michael immediately started receiving backlash from members in the field, including doctors, researchers, and other professors. Soon a campaign was started which urged Stony Brook University to force Michael to retire. Fortunately he still works at Stony Brook and was able to keep his job, even after so much backlash.

Guillermo Gonzalez was an astrophysicist professor at Iowa State University in Ames, Iowa. Gonzalez co-authored a book titled "The Privileged Planet," which made an argument that the universe was intelligently designed. Two years before Gonzalez was eligible for tenure at Iowa State, 130 faculty members at the university signed a petition that opposed "all attempts to represent intelligent design as a scientific endeavor." When he became eligible for tenure in April of 2007, Gonzalez's application was denied. While not specifically stated by the university, his position on intelligent design and dissent from his colleagues was most likely the cause.

The point being made here is evolution may be the only theory currently preached by scientists due to fear of ridicule from their peers or backlash from their academic institutions for even mentioning something else. There may be many more in academia who disagree with the theory of evolution, but from this same fear, they abstain from bringing up alternative theories and business goes on as usual.

This shouldn't be the case with any scientific

fields. All reasonable explanations should be accepted and studied by the scientific community, as this is how science progresses. We no longer believe bloodletting cures diseases or schizophrenia is caused from demonic possessions. Progressions in science has allowed us to learn more about life and the planet on which we live. Without competing theories, science stalls and we as a society stall. This should never be the case.

Evolution's Final Days

PART 9
Origin of Life

♦ ♦ ♦

Abiogenesis is the origin of life from non-living things. Throughout history, abiogenesis has constantly been disproven and has never been observed. Let's go back to a famous example you may be familiar with. Many years ago when humans left meat out, maggots would appear crawling out of the meat. At first it was theorized that the meat, a non-living thing, was actually creating the maggots through abiogenesis and people believed this for many years. A famous experiment in the late 1600's disproved this theory. When the meat was covered, thus preventing any creature from entering it, no maggots appeared. It was later discovered that flies were laying eggs in the meat, and this was the cause of the maggots, not abiogenesis.

Before humans, before dinosaurs, before primordial organisms, and before evolution was said to begin, we are told there was no life on this planet. The planet was just full of non-living gases, liquids, and solids, sloshing around the earth. Then, some sort of reaction occurred that brought about life spontaneously from non-living elements on earth. There is currently no generally accepted model for the origin of life, and scientists have never "created" life or a "protocell" in a laboratory. The most well-known explanation comes from the Miller-Urey "primordial soup" experiment in the early 1950's. In this experiment, two scientists sent a jolt of electricity through a flask of solution containing methane, ammonia, hydrogen and water; molecules which were thought to exist on early earth. This in turn resulted in the formation of around 9 amino acids. Amino acids are considered to be the "basic building blocks of life" as they are used in a number of biological functions. And this is where the textbooks leave off and then say "this is how life began and everything evolved from it… the end!" The issue here is we still do not know how life began. Okay amino acids, the basic building blocks of life have formed, now what? I have a picture of glycine, a common amino acid below.

If you've taken chemistry, you may recognize that an amino acid is simply an amine group (-NH2) and a carboxyl group (-COOH) attached to a carbon atom. And each amino acid has a side chain, which in glycine's case is hydrogen, which differs between amino acids. As shown in the photo, an amino acid is simply a few elements coming together to form a compound. Amino acids are NOT living things. They do not reproduce or replicate; they have no cells. A simple compound was created in this experiment, not biological life!

Going back to the 2nd law of thermodynamics, the universe gravitates towards disorder. So let us replace those bricks we threw out of the truck with these amino acids. Lightning hit the primordial soup, amino acids appeared, and then what happened? Based on what we know about the laws of nature, they would have become more disordered, spread out, and disassociated back into the soup. British astrophysicist Fred Hoyle has compared chemical reactions producing life,

> *"as equivalent to the possibility that a tornado sweeping through a junkyard might assemble a Boeing 747 from the materials therein."*

Again, abiogenesis has never been observed or conducted by anyone here on earth. Don't you think if it's a plausible hypothesis of how life formed, they would be able to do it in a laboratory? Or possibly we would see it happening occasionally here on earth in nature? Scientists currently have no explanation for what happened after this lightning strike. Based on what we know about earth, we can safely assume nothing would have happened, and life via abiogenesis was not the result.

PART 10
Effects of Mutation

◆ ◆ ◆

Different organisms, mammals, and plants here on earth have different DNA structures. Some parts of these structures are similar, however much of them vary from species to species. Considering this, the DNA had to somehow change and mutate in order to appear in its new form. First let's go over a brief summary of what DNA is. DNA is the genetic information we are born with and can be thought of as a blueprint your body uses to build itself. Your body looks to its DNA, which is found in the nucleus of cells, for how to construct your body. This would include features like your height, eye color, skin color, etc. DNA is made of four chemical bases: adenine, thymine, guanine, and cytosine. These bases pair with each other and form a phosphate wrapped double-helix structure, as shown in the picture below.

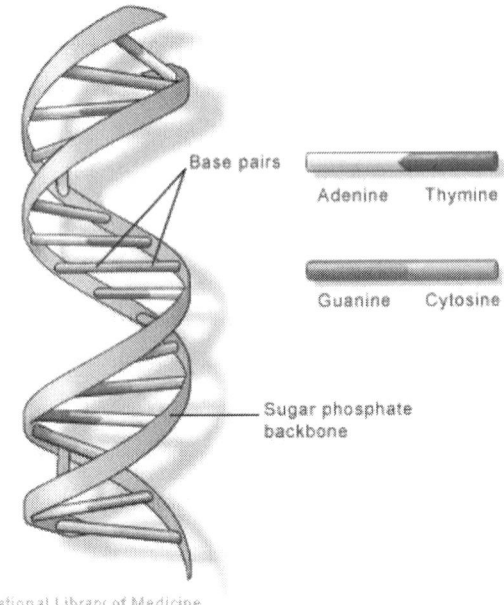

U.S. National Library of Medicine

There is a defined order to DNA, similar to a computer program. Examples being adenine only pairs with thymine, guanine only pairs with cytosine, and the helix of DNA only twists in a right handed formation.

Evolutionists claim small mutations in DNA over an imaginary time period is ultimately what has created the life we see on earth today. While this may be plausible for DNA of the same species as we see in nature, to change from a reptile to a mammal is in essence not possible. The issue is mutations in DNA cause negative effects, not positive. Positive effects of genetic mutation are not present in nature and evolutionists have failed to ever present an unequivocal, positive mutation that is purely positive with no negative consequences. Neutral, maybe. But positive? This has never been shown.

One example pointed to as a positive mutation is "Sickle Cell Anemia" which changes the shape of red blood cells in certain African populations. The benefit of this mutation is the blood of the affected human is more resistant to malaria, resulting in a survival advantage to that particular disease. But what about the negatives? If you've ever met anyone who has sickle cell anemia, it's a not a beneficial disorder to have. They have to take medication for their "positive" mutation, as without medical intervention their sickle-cell shaped blood cells would clog their arteries, resulting in lack of blood supply to their organs and ultimately death. Individuals who have this disorder and do not receive medical treatment tend to die at a young age. The only way these "beneficial mutations" are able to continue today is because of advanced medical intervention. Was medical intervention around 10,000,000 years ago to keep these "beneficial mutations" alive?

Charles Darwin was a scientist from the 1800's. He lived and died before modern medicine, before the discovery of DNA, and before our understanding of genetics. He believed in the inheritance of acquired characteristics, as in if a man lost his leg, he could have a son with a missing leg. And this is the man whom evolutionists base their entire theory on. This is important when you consider the human genome. The human genome is the total number of chemical bases found in the 23 chromosomal pairs contained in the nucleus. As in the amount of DNA contained in the 46 chromosomes human cells contain. A typical chromosome map is shown below, with the last chromosome pair determining the sex; XY being a male and XX being a female.

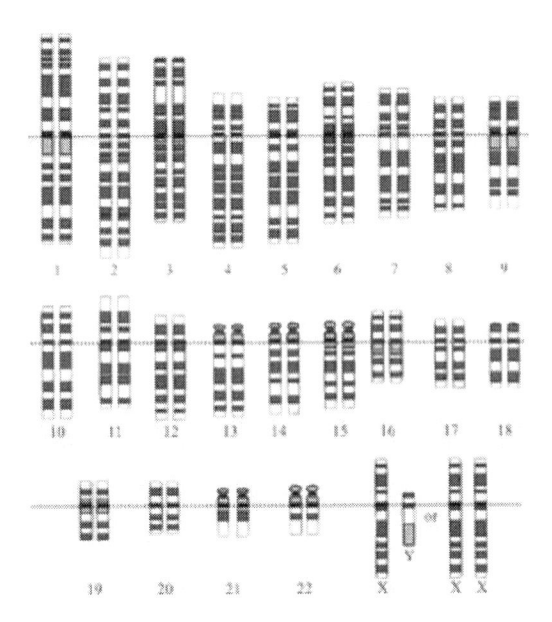

The human genome contains 6 billion base pairs

(A,T,G,C). So those 46 chromosomes shown in the picture above contain 6 billion base pairs inside of them. A random change of just three base pairs (0.0000000005%) is fatal. The lowest estimated difference between the DNA of apes and that of humans is 50 million base pairs... 50 million! Yet a change of just 3 of these base pairs is fatal. Unlike Darwin, we now have a much better understanding of how DNA works. And with this understanding, evolution starts looking more and more like a philosophical dream and less and less like science.

Human and Ape Chromosomes

Let's look at ape and human chromosomes. Apes have a total of 48 chromosomes while humans have a total of 46 chromosomes. Gametes are the sex cells of animals, and only contain half the chromosomes of normal cells, as a gamete from your mother (egg) and a gamete from your father (sperm) combine to form your full 46 chromosomes. The gametes of apes have 24 chromosomes while the gametes of humans have 23 chromosomes. That would mean somewhere along the way both the male and female apes had to lose one chromosome in their genome.

Four things can happen when we look at changes in chromosomes: a chromosome can be added, a chromosome can be deleted, a chromosome can be altered, and a chromosome can be moved. The issue is when we observe any of these changes in our current world, they produce only negative effects and often result in sterilization. This means hypothetically, even if the mutation was somehow beneficial, the mutant creature would not be able to pass on genes to their offspring due to being sterile.

I want to go over some examples of chromosomal changes that appear in humans, and the result of the chromosomal change. Humans with an extra copy of chromosome 21 are born with a disorder called Down syndrome. This results in mental retardation, organ defects, shorter lifespans, and a high chance of being born sterile. Humans with an extra copy of chromosome 18 are born with a disorder called Edward's syndrome which causes severe bodily defects, and a lifespan of not more than a few months. Humans with an altered 5th chromosome pair are born with a disorder named "Cri du chat," which translates to "Cry of the Cat." Children born with this disorder have severe mental retardation, a small head, and a cry that sounds like a distressed cat. Humans born with part of their chromosomes deleted (which would have had to occur between apes and humans) have severe birth defects and significant intellectual and physical disabilities. Humans born with a fragment of one of the 46 chromosomes moved to another location, known as translocation, can have problems with sperm or egg development, as in they become sterile.

We just went over a few of the chromosomal abnormalities that occur in humans. While there are more, you may notice they all have one thing in common... they produce negative outcomes! Nothing mentioned above, or in any of the chromosomal changes known produce "positive" effects. Even the so called "super-male" chromosomal change, where men are born with two "Y" chromosomes, is completely neutral. Most men who have the extra Y chromosome don't even know they are "super-males." There are no positive effects of chromosomal changes, but instead mostly negative effects, occasionally neutral, and many times these changes result in sterilization. This would mean nothing can be passed on to the next generation!

At the time Darwin was contemplating his theory, chromosomes weren't even discovered yet. The reason why individuals were born with Down syndrome or Edward's syndrome was unknown. To no fault of his own, just being born before modern times, Darwin had no understanding of the human genome. And yet we're supposed to take his theory, which is from the same time period arsenic was used to treat syphilis, as fact.

Customer Reviews

☆☆☆☆☆ 38
4.8 out of 5 stars

5 star	▬▬▬▬▬	87%
4 star	▪	10%
3 star		3%
2 star		0%
1 star		0%

See all 38 customer reviews ›

Share your thoughts with other customers

[Write a customer review]

If you are enjoying this book, could you please leave a review on Amazon? It would be greatly appreciated and allow me to come out with more informative books in the future. A shortened link to the review page is below:

fastlink.xyz/efd

Conclusion

♦ ♦ ♦

Evolution is an extremely flawed theory and our new understanding of genetics and science in modern times invalidates it. In terms of natural causes, scientists have nothing else to explain how this world came to be and thus are grasping for straws in the hope some evidence will eventually be found to help this theory. Faith is the word to use when describing the scientific community and the theory of evolution. Rebuttals made by the evolutionary community to the arguments made in this book are poor, and take a lot of outside the box thinking and extreme circumstances. They stray from logical reasoning and border on fringe science.

When we use our common sense, it becomes clear evolution as taught never happened. A better explanation for explaining life is an intelligent being designed humans and the other creatures that live beside us, and placed us on this planet. Whether that be the Christian-Judeo God, aliens, or some other type of being, that's up for you to decide. Scientists need to admit the theory of evolution has reached its breaking point and encourage their colleagues to come up with new theories that better explain how life on earth formed and continues to exist to this day.

I hope you enjoyed this book. Feel free to email me at john@unknownwealth.com with any questions or comments you have. I love to hear from my readers!

I plan on making more books in the future, that relate to science, the supernatural, and conspiracy theories. If interested in receiving a copy of one of my future books for free in exchange for an honest review, please email me with the subject line "Add Me to Your List John." I am the only one to ever have your email and will only send you messages that are related to new books.

If You Enjoyed This Book, You May Also Like...

The Chiropractor Hoax
The True Story of Chiropractic Medicine You've Never Been Told

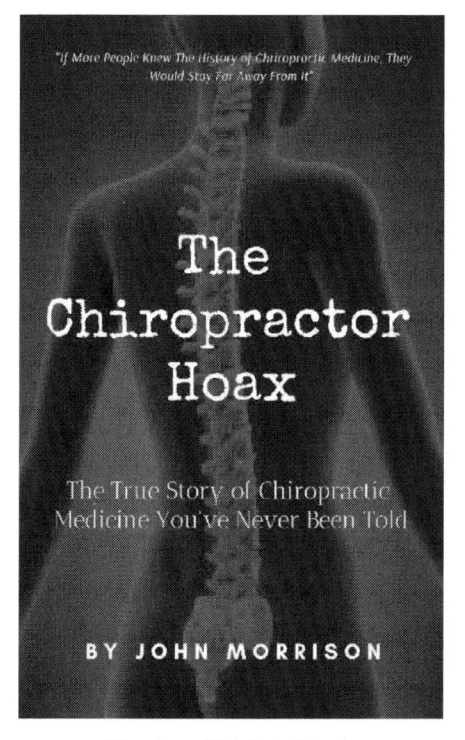

Shortened Link to Book
fastlink.xyz/chiro

Do not go to a chiropractor until you read this book! Chiropractic medicine is not what you think it is. In this tell all book, John Morrison breaks the lid on the chiropractic industry, and shows the unknowing public the truth behind the profession. Save your money and your health with The Chiropractor Hoax!

Printed in Great Britain
by Amazon